YOUNG ADULT SERVICES
STO

ACPL ITEM
DISCARDED
Y0-BVQ-655

Y 333.79 L17F   7113045
LAMBERT, MARK, 1946-
FUTURE SOURCES OF ENERGY

DO NOT REMOVE
CARDS FROM POCKET

ALLEN COUNTY PUBLIC LIBRARY
FORT WAYNE, INDIANA 46802

You may return this book to any agency, branch,
or bookmobile of the Allen County Public Library.

*Tomorrow's World*

# Future Sources of ENERGY

**Mark Lambert**

The Bookwright Press
New York   1986

# Tomorrow's World

Allen County Public Library
Ft. Wayne, Indiana

**Our Future in Space**
**The Telecommunications Revolution**
**Lasers in Action**
**The Robot Age**
**The Computerized Society**
**Living in the Future**
**Transportation in the Future**
**Future Sources of Energy**
**The Future for the Environment**
**Medicine in the Future**

First published in the United States in 1986 by
The Bookwright Press
387 Park Avenue South
New York, NY 10016

ISBN: 0-531-18077-8
Library of Congress Catalog Card Number: 85-73617

First published in 1986 by Wayland (Publishers) Limited
61 Western Road, Hove, East Sussex BN3 1JD, England

© Copyright 1986 Wayland (Publishers) Limited

Phototypeset by Kalligraphics Ltd., Redhill, Surrey
Printed in Italy by G. Canale & C.S.p.A., Turin

# Contents

7113045

Using energy 4

Energy from the sun 7

Earth, air and water 17

Energy from the ocean 27

Nuclear energy 33

Energy in tomorrow's world 43

Glossary 46
Further reading 47
Index 47

# Using energy

The word "energy" has been used in so many different ways in recent years, that it might be thought to mean many different things. Some foods are described as being "full of energy," and people speak of "energy resources" and an "energy crisis." We hear a great deal about nuclear energy and solar energy. In fact, energy has only one meaning – it is simply the capacity to do work.

For example, electrical energy is capable of making an electric motor turn. The chemical energy contained in an explosive can cause a bullet to be fired from a gun at high speed, and the moving, or kinetic, energy in wind can make a windmill turn.

The word "power" is often used to mean the same thing as energy – as in "nuclear power" and "solar power" – but, strictly, the scientific meaning of "power" is the rate at which work is done.

*Wind energy can be harnessed by a windmill.*

## Today's fuels

One of the most widely used forms of energy is electricity. At present, most of the world's electricity is generated using fossil fuels – coal, gas and oil. When they are burned, sufficient heat is released to make steam, which is used to power huge steam turbines. These turn generators to produce electricity. Fossil fuels are also used in other ways. Among the products of an oil refinery are kerosene, gasoline and diesel – all vital fuels for transportation. Oil products also include a wide range of chemicals that can be converted into plastics, synthetic fibers, drugs, insecticides and explosives. The petrochemical industry is so important in the modern world that some people think that oil is too precious to waste as a fuel.

*Large electrical generators driven by steamturbines.*

*An Oil Refinery.*

Coal is burned as fuel and also produces chemicals. Other fuels include natural gas, bottled gas (a natural gas product) and wood. All these fuels, except wood, are non-renewable, and supplies will eventually be exhausted. Even wood, which is used extensively in Third World countries, cannot be replaced as fast as it can be burned. Nuclear power may solve some problems but it creates others. Fortunately, we are surrounded by a variety of forms of energy. For the future, people are looking for new ways of producing energy – in particular, ways of harnessing the vast amounts available from the sun, wind, earth and sea.

# Energy from the sun

Every day, large amounts of heat and light from the sun reach the earth's surface. Much of the heat contributes toward keeping the planet warm enough to support life. Plants use some of the light in their vital food-making process, which is called photosynthesis but, even so, there is plenty of heat and light left for us to use as a source of energy.

Solar heating panel set into a roof in Australia.

The best-known heat collector is the flat plate collector, or solar panel. This type of collector is basically a radiator working in reverse. It usually consists of a number of fine tubes in or on a heat-absorbing material. The panel is painted black – the color that absorbs most heat. To help trap heat and keep out rain, the panel usually has a glass cover. Water pumped through the panel heats up and passes to a hot water tank, where it heats a hot water supply. Except on very hot days, solar panels can seldom do much more than provide enough heat to warm a domestic water supply. But they can help to reduce the cost of heating water by other means.

Another type of solar collector is the air collector. Air replaces water in the tubes, and the heated air is driven by a fan into the living space of the house. As with the flat plate water collector, this can produce only moderate amounts of heat.

A vacuum-tube collector can generate higher water temperatures of 90°C–300°C (300°F–575°F). Various designs are available but, basically, the heat-absorbing fluid is contained in a tube that is separated from the outside air by a vacuum. This reduces the heat losses due to conduction (the transmission of heat through a substance, caused by kinetic energy moving the atoms), and convection (the passage of heat in a gas or liquid by means of moving currents).

Similar temperatures can be achieved using collectors that focus the sun's rays. There are various designs today. Some have parabolic (curved) reflectors, which focus the rays onto heat-absorbing tubes in which liquid is flowing. The rays are automatically focused onto the tube, regard-

*Curved reflectors focus the sun's rays onto heat-absorbing tubes.*

*At Odeilo, in France, sixty-three sun-tracking flat mirrors concentrate the sun's rays on to a huge curved reflector.*

less of the sun's position in the sky. Other systems track the sun. Flat mirrors, which are easier to make than curved mirrors, can be used in tracking systems. Yet another system uses a special type of plastic lens, known as a Fresnel lens, instead of mirrors.

The highest temperatures – over 300°C (575°F) – are achieved using systems that focus the sun's rays onto a single point rather than a long tube. One type, known as a parabolic dish reflector, looks rather like the dish of a radio telescope or a satellite tracking antenna. This type of collector can achieve temperatures of up to 900°C (1,650°F). A smaller, portable version of this has been developed as a solar oven. The world's largest parabolic reflector forms part of the solar furnace at Odeilo in France. This has an array of sixty-three, sun-tracking flat mirrors to concentrate the rays from a large area onto the curved reflector, which in turn focuses the rays onto a furnace. At the solar plant in Odeilo, temperatures of over 3,000°C (5,400°F) can be achieved.

Systems that employ flat mirrors to collect the sun's rays over a large area are known as central receiver collectors. There may be hundreds or even thousands of mirrors, each of which directs the sun's rays toward a point near the top of a thermal tower. A liquid – oil, high-pressure water, molten salt or liquid metal – flows past the focus point and carries the heat away to electricity-generating equipment. The largest solar power plant is near Barstow in California. It has 1,818 mirrors arranged in circles around the tower, and can generate up to 10,000 kilowatts of electricity. Future solar heat power stations should be able to generate up to 30,000 kilowatts.

*A solar power plant. Arrays of flat mirrors arranged in a circle focus sunlight onto a central tower, where a liquid or gas is heated to provide vapor.*

# Storing the sun's heat

One of the problems of solar heating is that the sun does not always shine when heat is needed, particularly at night and during the winter months. To overcome this, various ways of storing heat have been devised.

A solar pond is a collector and storage system combined. It has a dark, heat-absorbing lining at the bottom, and is filled with layers of brine (salt solution), each layer being less concentrated than the one below. Heat from the sun is absorbed by the dark lining and passed to the most concentrated layer of brine. The heat is prevented from escaping by the less concentrated layers above, and the hot brine can be piped to where heat is needed. Solar ponds are cheap to use and can be built on a large scale.

*A solar pond in the Dead Sea area collects and stores heat energy.*

*A parabolic dish reflector focuses the sun's rays onto a single point.*

Heat can also be stored for some time in water tanks if they are placed underground and are well insulated. Another method of storing heat is being tried out in a low-energy housing project in Sweden. On a sunny day, sunlight penetrates a glass-covered street, and the heat is used to heat up water, which is then pumped into underground storage rocks. On cool days, the process is reversed, the rocks giving up their stored heat to water, which is pumped to the houses. On a small scale, heat can also be stored in rock beds constructed underneath houses.

Chemicals, too, can be used to store heat. Large amounts of heat are taken up by a substance when it melts. Heat is given off again when that substance solidifies. Some systems make use of this process.

## Using the sun's light

Energy from the sun can be stored as electricity. Solar cells, or photovoltaic cells, convert light energy directly into electrical energy. Solar cells can be made using several different kinds of materials known as semiconductors, but the most commonly used material is silicon.

Individual cells generate very little power, so large arrays of cells are needed for most purposes. Panels of solar cells are already being used to power satellites, unmanned lighthouses, aircraft navigation beacons, telephone booths in remote places, and even very lightweight airplanes. The cost of solar cells is decreasing rapidly, and the efficiency of solar arrays is being increased. In some parts of the

*Large array of photovoltaic cells, used to convert light energy directly into electrical energy.*

*Solar cells enable telephone links to be established in remote places.*

world which have long hours of sunshine, such as California, Egypt, Pakistan and Saudi Arabia, there are houses and villages or towns equipped with solar arrays to supply electricity. In California, there are now several solar power stations capable of generating over 3,000 kilowatts of electricity, using rows of sun-tracking solar arrays.

One energy scheme for the future envisages huge solar arrays in space. An array capable of generating 5 million kilowatts of power would have 10 million solar cells on a frame covering 50 sq km (19 sq miles). The energy collected by such an array would be transmitted from a huge antenna, 1 km across, in the form of microwaves. On the ground 36,000 km (22,500 miles) below, the receiver would cover an area of 130 sq km (50 sq miles). Such solar power stations would not have their energy source filtered by the earth's atmosphere or have it cut off by clouds. However, the microwave radiation could cause problems, and the cost of constructing solar power stations in space is likely to be enormous.

*Impression of a huge solar array in space that would transmit energy in the form of microwaves to a huge antenna on earth.*

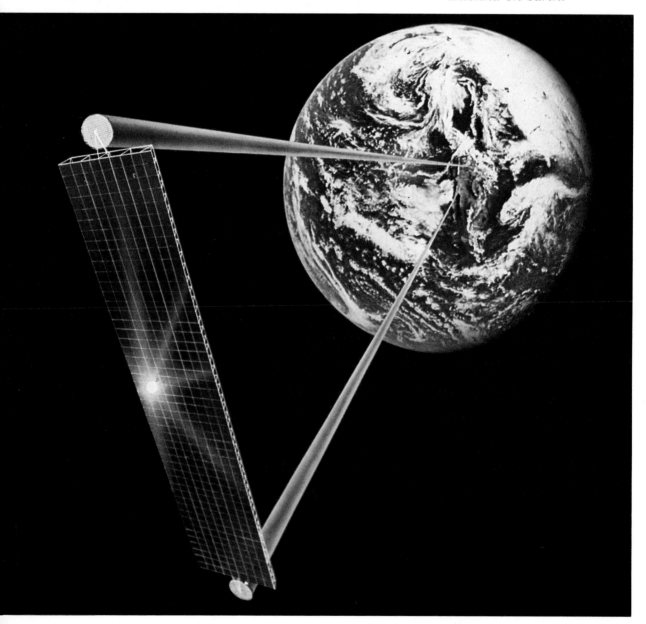

# Earth, air and water

The modern industrial world could not exist without large amounts of energy. Beneath the ground there are supplies of oil and coal. But long after these run out, there will be underground heat which can be used. Air also contains a small amount of heat energy, and the vast amounts of kinetic energy present in wind and waterfalls can be harnessed effectively.

## Heat from underground

The earth's crust is a thin layer of rock, about 30 km (19 miles) thick. In places, the temperature at this depth may be over 1,000°C (1,800°F). Such temperatures are maintained by the heat given off as radioactive atoms inside the earth decay. Rocks deep in the earth's crust are therefore very hot. In some places, rainwater has seeped down into hot sedimentary rock several kilometers below the surface. Where the rock is very porous or broken, the water may return to the surface as hot springs or geysers. These often occur where the crust rock is thin, or where there is volcanic activity, and molten rock comes near the surface.

*Geothermal plant.*

*A geothermal power station in New Zealand.*

In other places, hot underground water supplies, or aquifiers, lying one or two kilometers below the surface, can be reached by drilling. If the temperature of the water is very high, it comes to the surface under pressure, and can be used to generate electricity. Alternatively, the hot water can be pumped to the surface and used to supply heat to homes and factories. The used water is usually pumped back down the well via a second borehole. Some of the best geothermal ("earth heat") areas occur in Italy, the United States and Japan.

Water heated naturally by geothermal energy in sedimentary rocks is already being used in many parts of the world. However, impermeable underground rocks, such as granite, contain even more heat. Engineers are currently working on a way to extract this heat. Basically it is done by causing the hot, dry rock to break up and then passing water through the broken rock.

First, the rock is drilled; the bore may be two or three kilometers deep. Then explosives can be used to break up an area of rock at the bottom of the bore. Water pumped down under pressure helps to extend the area of broken rock. A second bore is drilled so that the water, now heated, can flow back up to the surface.

Unlike solar, tidal and wind energy, geothermal energy is not renewable; a geothermal well can only be expected to produce useful heat for between thirty and fifty years. However, geothermal energy is cheap and easy to produce without having any undesirable effects on the environment. When a well does "run out" of heat, a new well can be drilled within 1,000–1,500 meters (1,100 yds–1,640 yds) of the original bore.

In the future, it may become possible to exploit geothermal energy at its source. In the U.S. and the U.S.S.R., engineers are experimenting with ways of extracting heat from molten volcanic rock, and if water is injected into molten rock under the right conditions, fuel gases such as hydrogen and methane can be produced.

*Engineers may be able to extract heat from molten volcanic rock in the future.*

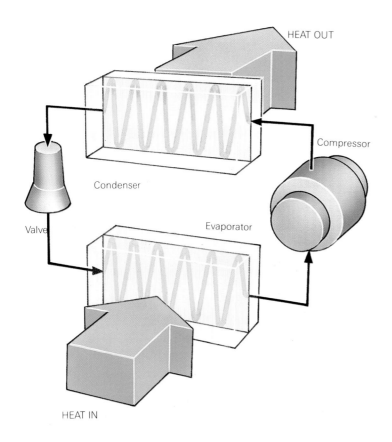

*Diagram of how a heat pump works.*

## Heat pumps

Solar collectors and geothermal installations often produce heat at relatively low temperatures. A heat pump can be used to "pump" this heat up to a higher temperature. A heat pump works like a refrigerator in reverse. The low temperature of the heat source is used to cause a liquid in a low-pressure container to evaporate. The vapor is then pumped and compressed by a compressor (a device, such as a pump or turbine, used to increase the pressure of a gas by squeezing it into a smaller space) to a high pressure container, where it condenses, giving out its heat.

A heat pump can use waste heat from a machine or from the drainage system of a house. It can be used to upgrade the heat produced by a solar collector or geothermal well. The useful life of a geothermal well can be doubled in this way. Even the air outside a house or the water in a stream can be used as sources of low temperature heat.

## Wind energy

Anyone who has experienced being out in a gale can appreciate how much energy is contained in wind. This can be harnessed. In fact, people have been using wind energy for thousands of years – for milling, grinding and pumping, and in sailing ships. But only recently has wind energy come to be used to generate electricity.

*A modern horizontal-axis wind turbine.*

There are two main types of modern windmills, or wind turbines. There are those that, like traditional windmills, have horizontal axes, which means that the wind strikes blades positioned in the same plane as the upright tower or post, causing them to rotate in the "conventional" way. Whereas traditional windmills have four or more blades, modern wind turbines have only three or even two blades. In most cases the blades act like airfoils. Small turbines can rotate at very high speeds. Large turbines rotate more slowly and need stepping gears to operate electricity generators.

The main problem with wind is that its strength is not constant. Some wind turbines are designed to operate only

*A vertical-axis wind turbine.*

*Altamont Pass wind farm, California, where 300 wind turbines generate electricity.*

between certain wind speeds. Others have blades that can be "feathered" so that the pitch (the angle at which the blade meets the air) can be altered, allowing the turbine to rotate at a constant speed whatever the wind force. Such turbines can continue to produce power even in a gale.

The second type of wind turbine has a verticle axis. This is not a new idea. Such windmills date from about 600 A.D.

*Arrays of Musgrove wind turbines at sea. They can alter their shape to suit the speed of the wind.*

and may be even older than horizontal-axis machines. As the wind strikes the blades, which are positioned at right-angles to the central tower or post, they rotate rather like an eggbeater. Vertical-axis machines have several advantages. They can react to wind in any direction, and so do not need equipment to turn them into the wind. They also require less support. One recent type of vertical-axis wind turbine, the Musgrove turbine, alters its shape according to the wind speed.

There are already a large number of wind turbines in use. Some are capable of generating 3,000 kilowatts of electricity. Problems that have yet to be fully overcome include overspeeding (owing to failure of control devices), blade fatigue and failure of electrical equipment. But in the future, it is likely that more and more electricity will be generated using wind power. Some of the most suitable sites are at sea, where winds are strongest and most reliable. Within the next thirty years, there may be arrays of wind turbines constructed on offshore platforms.

## Hydroelectric power

Where water falls from one level to another, its kinetic, or moving, energy can be used to power electricity-generating turbines. The power that can be extracted from a hydroelectric project depends on the rate of flow of the water and the head – the height of the water above the turbines. Some hydroelectric power stations are situated at waterfalls, where the head is naturally high. Other plants use dams to raise the water level. The largest hydroelectric project, the Snowy Mountain system in Australia, uses

*At some hydroelectric power stations, dams are used to raise the water level.*

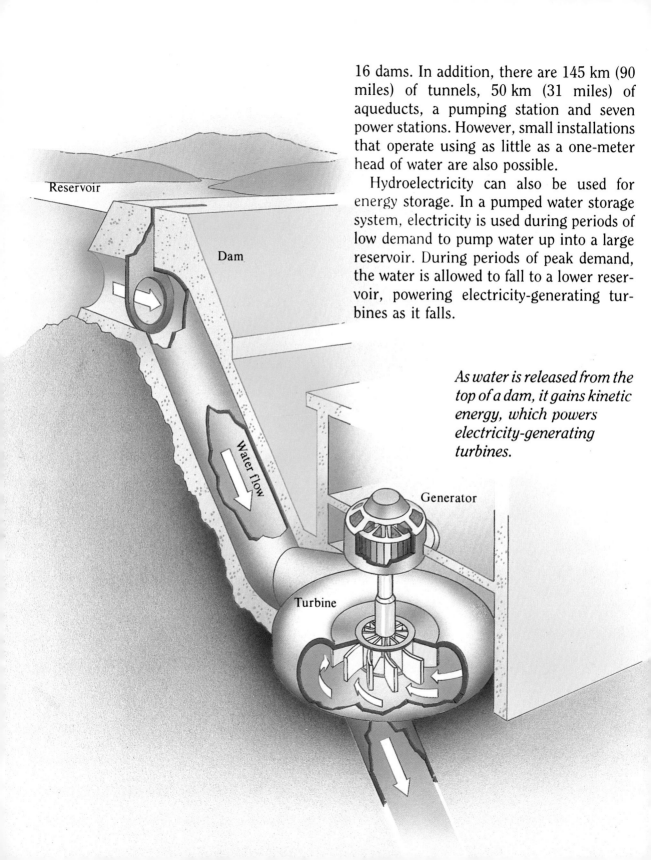

16 dams. In addition, there are 145 km (90 miles) of tunnels, 50 km (31 miles) of aqueducts, a pumping station and seven power stations. However, small installations that operate using as little as a one-meter head of water are also possible.

Hydroelectricity can also be used for energy storage. In a pumped water storage system, electricity is used during periods of low demand to pump water up into a large reservoir. During periods of peak demand, the water is allowed to fall to a lower reservoir, powering electricity-generating turbines as it falls.

*As water is released from the top of a dam, it gains kinetic energy, which powers electricity-generating turbines.*

# Energy from the ocean

The world's oceans are constantly moving. Currents, tides and waves provide us with a potential source of large amounts of kinetic energy, and heat energy is available in tropical seas – if only it can be harnessed economically.

## Tidal power

Every day, the combined effect of the moon's gravity, the sun's gravity, and the rotation of the earth on its axis, causes vast quantities of water to move around the world's seas. We see these movements of water as tides. At sea the rise and fall of the tides is only about 1 meter (39 inches), but along coasts, tides combine with currents and the shape of the land to produce much greater rises and falls.

The best-known tidal energy project is in the Rance estuary near St. Malo in France. It was built in 1966 after years of research. It consists of a single barrage, (a dam), across the mouth of the estuary. Inside the barrage are twenty-four turbines that can produce a total of 240,000 kilowatts of electricity, using the ebb and flow of the tide.

Tidal projects are in use in the U.S.S.R., China and Canada. Engineers in other countries are currently studying ways of harnessing tidal energy. But tidal power stations are expensive and do not necessarily produce electricity when it is most needed. Further, such projects may have undesirable effects on the animals and plants of the estuary.

## Wave energy

Ever since the oil crisis of 1973, people have been looking for ways of turning the constant rising and falling motion of waves into useful energy. In some places waves can be up to 25 meters (82 feet) high and may contain 50–70 kilowatts of power per meter of wavefront. Even if only a third of this power can be converted into energy usable on the mainland, a future wave energy power station may become feasible.

*Rows of Salter ducks could convert wave motion to electrical energy.*

Over 300 different designs for wave energy converters have been patented. In many of these, the up-and-down movement of a device floating in the water is used to operate one or more hydraulic pumps, which pump fluid to a turbine-driven generator. An early design was the Salter duck, devised by Stephen Salter of Edinburgh University, Scotland. The Salter duck is still being refined and improved. Another promising and simple design is the Bristol oscillating cylinder.

A different range of devices is based on the oscillating water column (OWC) principle, in which a column of water in a box with an underwater opening acts like a piston in a cylinder. As the water column moves up and down, air

*The oscillating water column. The incoming wave raises the level of the water in the column, squeezing the air above it and pushing it through the valve, where it drives an air turbine and generates electricity.*

*The water level drops in the chamber, sucking air from above. The air drives the turbine on its way to the chamber, generating electricity.*

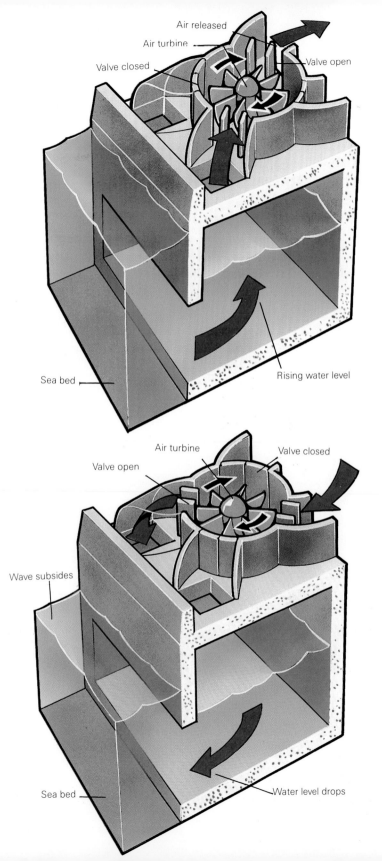

is forced through a turbine, using a system of one-way valves. The first OWC device was devised in Japan to power a light on a navigational buoy. Since then much research has been carried out in both Japan and Britain. Air pressure is also used in other devices, such as the Lancaster flexible air bag and the "sea clam."

One of the main problems of wave energy converters is that in order to generate large amounts of power they must be very large. Anchoring such devices will not be easy. To overcome this, the National Engineering Laboratory in Scotland has developed a breakwater oscillating column device that would stand on the bottom in water 15–20 meters (49–66 feet) deep. But in spite of this and all the other projects that have been proposed, it will be a long time before wave energy converters become a practical proposition. As yet, no one knows what environmental effects might occur as a result of using large devices that would dramatically alter the wave patterns reaching the shore.

*The sea clam is a wave energy converter which uses air pressure.*

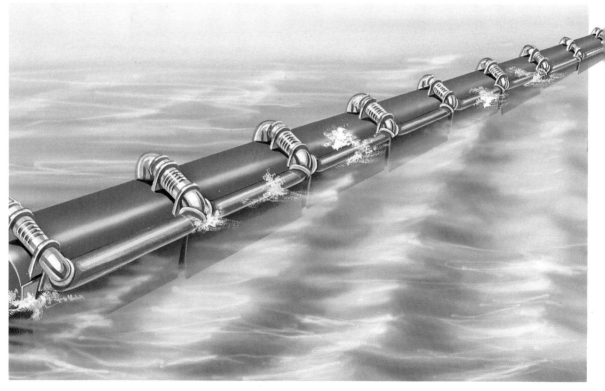

## Heat from the sea

The idea of extracting heat from the sea was first conceived over 100 years ago. The first Ocean Thermal Energy Conversion (OTEC) project was devised in the 1920s.

An OTEC plant takes advantage of the fact that in tropical regions, the upper, solar-heated layers of the sea are much warmer than the deeper layers. At the equator, for example, the temperature changes from 26°C (79°F) in the surface layer of the water, to about 8°C (46°F) at a depth of 450 meters (1,475 feet). In an OTEC plant, warm water from the upper layer is used to evaporate a liquid with a low boiling point, such as ammonia or freon. As with steam in a conventional power station, the vapor produced is then used to drive a turbine, which turns a generator, producing electricity. Once the vapor has done its work, it is condensed in a condenser cooled by water drawn up from below. The liquid ammonia or freon is then pumped back to the evaporator and re-used.

## Future OTEC projects

Most of the OTEC projects proposed so far are designed to float on the ocean. Among the largest are two suggested during the 1970s by TRW Inc and Lockheed Missiles, both of California. The TRW design is a massive concrete cylinder extending down 1,200 meters (3,937 feet) into the sea, and topped by a platform above the surface. The Lockheed design has a telescopic pipe, made of five sections, which would collect cold water from a depth of 500 meters (1,541 feet). All that would show at the surface would be the tip, or "spar buoys" of the power plant. This would make the power plant less vulnerable to tropical hurricanes.

Although some small-scale experiments have been carried out, no large OTEC plant has yet been built. The largest is a 100-killowatt, experimental plant in Japan. This is a land-based installation that has warm and cold water pipes extending into the sea to appropriate depths. A similar plant is being built in Bali. Investigations are being made into the effect of OTEC projects on marine life.

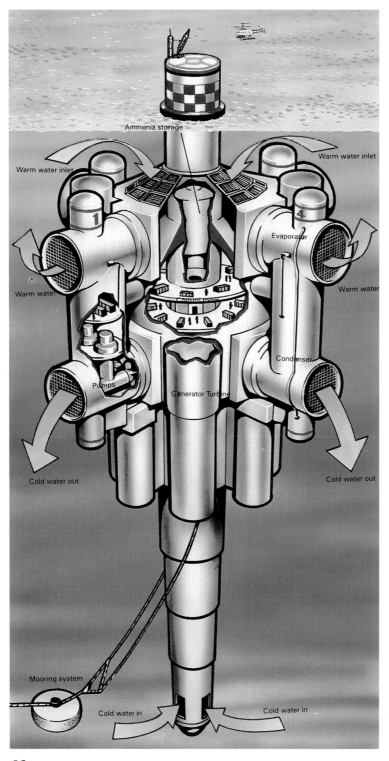

*An OTEC plant uses the varying temperatures present in the ocean to convert liquid to vapor (which powers turbines), and then reconvert it to liquid.*

# Nuclear energy

All matter consists of tiny particles called atoms. Each atom is made up of a nucleus, containing protons and neutrons, surrounded by electrons. Normally, atoms cannot be divided, but the atoms of certain substances are unstable and decay into smaller atoms, giving out radiation in the process. Such elements are described as radioactive. One such element is uranium, and in 1939 scientists discovered that if a certain type of uranium was bombarded with neutrons, its atoms split, releasing vast amounts of energy. This discovery led to the building of the first atomic bomb. It also led to the development of the nuclear power industry.

*A Uranium-235 atom absorbs a stray neutron and undergoes fission.*

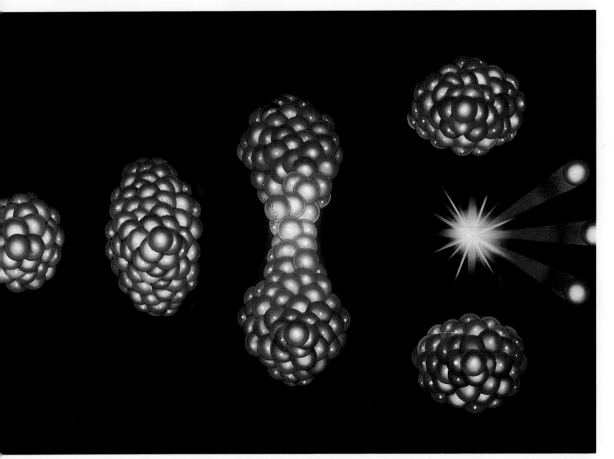

# Nuclear fuel

In its natural state, uranium consists of two different forms, or isotopes – uranium-235 and uranium-238. Uranium-235 atoms contain three fewer protons than uranium-238 atoms and are less stable. If a uranium-235 atom absorbs a stray neutron, it splits up (undergoes fission) into two smaller atoms and more neutrons. These neutrons can go on to strike more uranium-235 atoms, producing a series of chain reactions.

If the chain reactions are allowed to proceed uncontrolled in a fairly concentrated amount of uranium-235, the result is a nuclear explosion. In practice, uranium-235 exists in tiny amounts, together with much larger amounts of uranium-238, which does not undergo fission. To make nuclear fuel, uranium is often enriched to increase the uranium-235 content to between two and three percent.

*Fuel subassembly being lowered into the core tank.*

*Dounreay Nuclear Power Station in Scotland, with a prototype Fast Breeder Reactor.*

## Using nuclear power

The heat produced by a controlled nuclear-fission chain reaction can be used to generate electricity. Inside a nuclear reactor, the core contains nuclear fuel in fuel rods. Surrounding these is a material called the moderator, which slows down the neutrons and makes them more efficient at splitting uranium-235 atoms. Also in the core are control rods, which contain a neutron-absorbing material. The control rods can be raised to allow the chain reactions to proceed more quickly, or lowered to slow down or stop the process completely. To remove the heat produced in the fuel rods, a cooling fluid is pumped through the core. Outside the core, the cooling fluid is used to generate steam, which powers turbine-driven generators.

*When a Uranium-238 atom is struck by neutrons, plutonium-239 atoms are formed.*

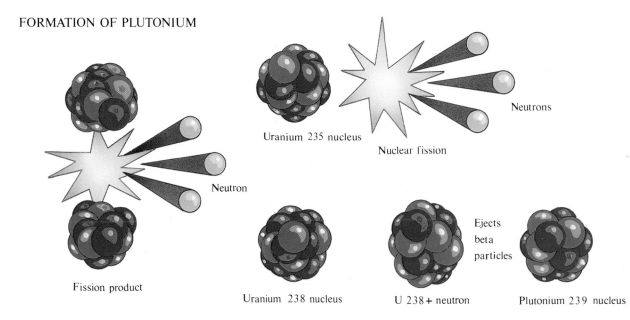

FORMATION OF PLUTONIUM

Uranium 235 nucleus — Nuclear fission — Neutrons — Neutron — Fission product — Uranium 238 nucleus — U 238 + neutron — Ejects beta particles — Plutonium 239 nucleus

There are various designs of nuclear reactors. In the United States engineers have preferred water-cooled reactors. In the Pressurized Water Reactor, water not only acts as the coolant but also as the moderator, and in the Boiling Water Reactor the water is turned to steam and used to power a steam turbine directly.

In Britain, however, engineers have so far favored reactors cooled by gas, such as the Magnox and Advanced Gas-Cooled Reactors. The latest type is the High Temperature Gas-Cooled Reactor, which uses helium as the cooling gas. The fuel is highly-enriched uranium oxide, which is made in the form of tiny spheres. These are embedded in graphite, which acts as the moderator.

Among the latest reactors are Fast Breeder Reactors, which can actually make more fuel than they use. A Fast Breeder Reactor contains no moderator, so the neutrons are not slowed down. However, when a fast-moving neutron does manage to split a uranium-235, more new neutrons are produced and this increases the rate of the chain reaction. In addition, when fast-moving neutrons strike uranium-238 atoms, they produce plutonium-239 atoms. Plutonium-239 is itself a fissionable material and is therefore a nuclear fuel. Eventually a point is reached at which more plutonium-239 is being produced than is being used.

*This diagram shows how a Gas-Cooled Magnox Reactors works.*

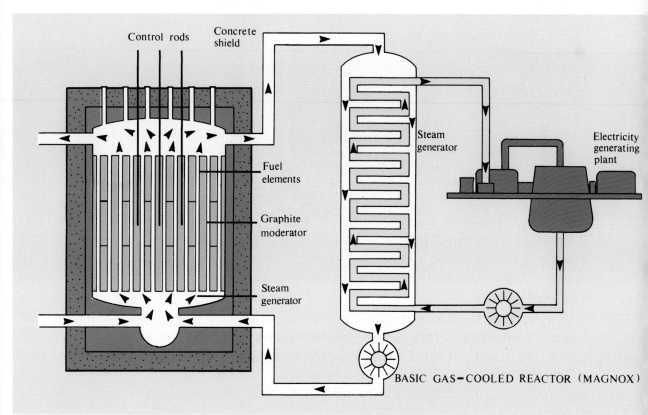

## The Nuclear Age

We live in what is often known as the Nuclear Age. The future of the nuclear industry is not, however, at all certain. Technically, there is no reason why nuclear power should not supply electricity to a large part of the world for a long time to come. Fast Breeder Reactors could increase the world reserves of nuclear fuel up to sixty times.

*The problem of the disposal of nuclear waste might be partially solved by burying it deep in the earth.*

Other factors must be considered. Nuclear power stations produce dangerous radiation, and over the last thirty years several serious accidents have occurred. The cost of meeting the latest safety regulations is high, and this has contributed to the decline of the world's nuclear industry over the last few years.

The main reason for the decline, however, is simply that nuclear power is becoming unpopular. Those in favor of the future use of nuclear power argue that existing safety measures will prevent any major disaster from occurring, but there are many who believe the risks are too high. The unsolved problem of what to do with radioactive waste still remains. Plans to store it deep underground or at sea have

*Deuterium and tritium pellets fuse to form a helium nucleus and one neutron.*

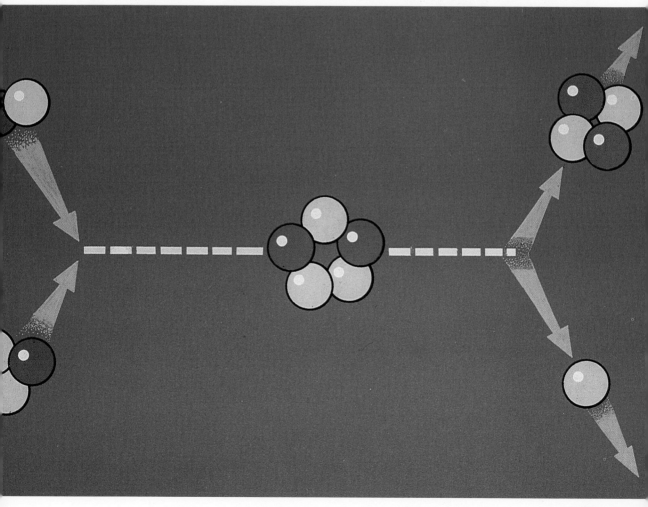

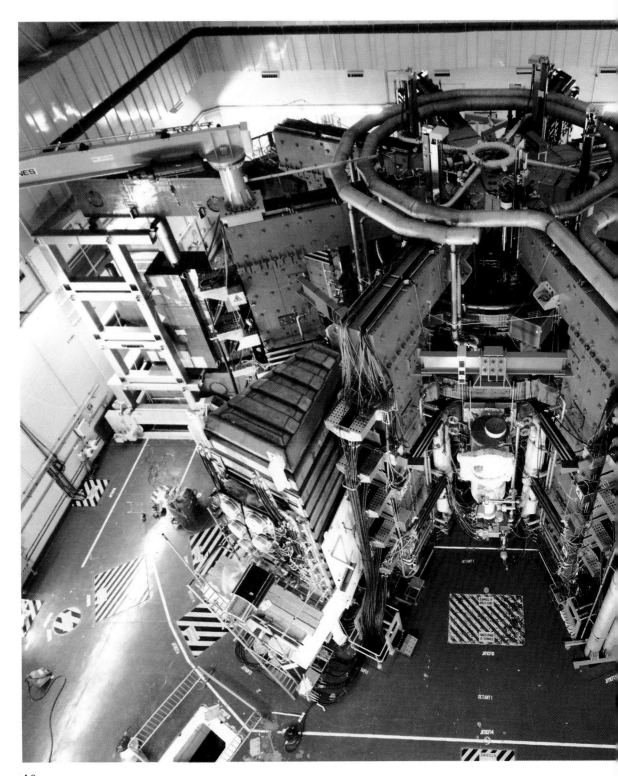

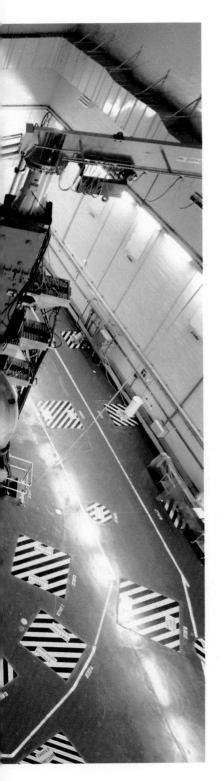

been strongly opposed. Finally, there is the fact that nuclear fuel can also be used to make devastatingly destructive weapons. Some people believe that no security system can guarantee the safety of nuclear fuel.

## Nuclear fusion

In nuclear fission, atomic nuclei are split into several pieces. Nuclear fusion, on the other hand, involves forcing two atomic nuclei together to make one nucleus. This process produces vast amounts of energy, but unlike nuclear fission it produces almost no harmful radiation.

Nuclear fusion is the process that powers the sun. Here on earth it could supply almost unlimited amounts of energy. The process involves fusing two forms of hydrogen – deuterium and tritium – to form helium. However, this can only occur when the two kinds of hydrogen are squeezed together at very high temperatures. The temperature inside the sun is 15,000,000°C (27,000,000°F), and even this temperature is too low to produce useful power here on earth.

Much research into nuclear fusion is being carried out. At the very high temperatures which are necessary, the materials involved take the form of plasmas (very hot gases). The problem is to contain such plasmas and squeeze them for long enough to produce useful energy. Today the most promising design of fusion reactor is the tokamak design, of which there are a number of versions. Basically, the plasma is both created and held by magnetic coils arranged in a circular tube, like the inner tube of a car tire.

Another design of fusion reactor consists of a round chamber surrounded by a number of very powerful lasers. The aim is to use laser beams to implode (explode inwards) tiny pellets of deuterium and tritium. However, as with

*Tokamaks contain and squeeze the very hot plasmas produced when pellets of deuterium and tritium undergo fusion.*

Tokamak reactors, this process requires the input of large amounts of energy and, as yet, no fusion process has been able to produce more power than is put in. Despite all the work, there is no certainty that nuclear fusion will ever become a reality.

*NOVA – the world's most powerful laser – is used in the process of nuclear fusion.*

# Energy in tomorrow's world

In the last ten years or so we have heard a great deal about the world's "energy crisis," but no such crisis has occurred. The world is not about to run out of available energy. Nor do we lack the technology to make good use of "alternative," renewable energy sources, such as solar power, wind power and wave power. What is true, however, is that ever since the Arab–Israeli war of 1973 triggered the first oil crisis, we have had to accept the fact that oil is not a cheap, unlimited source of energy.

The future of energy sources is not easy to predict. Enthusiasts of the various forms of alternative energy confidently predict that oil will be replaced by one or more of these forms of energy in the next twenty to thirty years. It is unlikely that the fossil fuels, which currently meet 88 percent of the Western world's energy needs, will cease to do so suddenly. Oil supplies may well last considerably longer than predicted. New oil fields are being discovered every year, and even when they do run out, processes already exist for turning coal into oil. Coal supplies are likely to last for over 250 years.

*In Pakistan, solar electric pumps are replacing the animal-powered water-lifting devices.*

## Small- and medium-sized energy projects

On the other hand, other sources of energy will certainly be developed. One of the advantages of alternative energy sources is that they can be successful on a small scale.

The attractions of small-scale projects are that they are cheap, and are not controlled by the government or large, multinational companies. Small-scale domestic projects, employing solar power or wind power, and backed up by energy conversion measures, work well, particularly if assisted by larger-scale projects such as local coal-fired power stations.

*Wind farms may provide energy in the future.*

## Fuels

A wide range of fuels may become economically possible in the future. Biogas (mostly methane) can be produced from garbage, sewage and animal waste. Wood can be burned as a fuel, or used to produce gas or methanol.

In the future, hydrogen could become one of the world's major fuels. It can be produced by splitting water into hydrogen and oxygen in a process known as electrolysis, which requires electricity. Scientists are trying to develop chemical processes that mimic photosynthesis – the process by which plants produce food and oxygen.

## Large-scale energy projects

Political and economic considerations will play a major part in deciding which of the various potential energy sources are developed.

At present, it looks as though wind energy could become the most widely used alternative energy source in the northern hemisphere, whereas in tropical and subtropical regions, the use of solar power will probably increase. Geothermal projects are being studied in many parts of the world. Tidal and hydroelectric projects are also being studied, but many of these seem likely to have undesirable effects on the environment. Wave energy and OTEC projects seem promising, but, at present, only Japan is seriously considering the possibility of using wave power.

Energy in the future is therefore likely to come from a variety of sources. As reserves of oil are used up, coal and nuclear power may provide more and more of the world's energy needs. On the other hand, it is possible that fossil fuels may be reserved for the production of chemicals, and that nuclear power may one day be abandoned. In this case, all the world's energy will come from renewable sources.

*A solar home in New Mexico: a windmill pumps water heated by solar panels.*

# Glossary

**Atom** The smallest quantity of an element that can take part in a chemical reaction.
**Chemical energy** Energy in a substance resulting from the arrangement of its atoms.
**Condenser** A device in which a gas or vapor is cooled to a point at which it becomes liquid.
**Electrical energy.** Electricity. Energy created by the flow of free electrons in a substance.
**Electrons** Negatively-charged particles that move around the nucleus of an atom.
**Energy** The capacity to do work, such as making an object move or raising its temperature.
**Evaporate** To change or cause to change from a liquid or a solid into a vapor, or gas.
**Fissionable** Capable of undergoing nuclear fission – the process of splitting an atomic nucleus into smaller particles.
**Fossil fuels** Any natural fuel, such as coal, oil or gas, formed underground from the bodies and tissues of prehistoric animals and plants.
**Generator** A device that produces electricity by causing one or more coils of wire to rotate in a magnetic field.
**Harness** To capture and control the energy contained in something.
**Heat energy** 1. A form of kinetic energy created by the rapid vibration of the atoms or molecules of a substance. 2. Infrared rays, a form of radiant energy.
**Isotopes** Atoms of the same element that have the same number of protons and electrons but different numbers of neutrons.
**Kilowatt (KW)** 1,000 watts. The watt (W) is the basic standard unit of electric power, 1,000,000 watts (1,000 KW) = 1 megawatt.
**Kinetic energy** Energy due to movement.
**Mechanical energy** A form of kinetic energy in which the movement is performed by a mechanical device.
**Neutron** An uncharged particle in the nucleus of an atom.
**Nuclear** Pertaining or relating to a nucleus, the central part of an atom.
**Nuclear energy** Energy released through the alteration of atomic nuclei.
**Nucleus** The central part of an atom, consisting of protons and neutrons.
**Potential energy** Energy in an object due to its relative position.
**Proton** A positively-charged particle in the nucleus of an atom.
**Radiant energy** Energy that can cross space, such as light, infrared rays and microwaves.
**Radioactive** Having atoms that disintegrate of their own accord, releasing smaller particles and sometimes harmful rays.
**Thermal** Concerned with heat.
**Turbine** A device powered by steam, gas or water, which forces two or more blades to revolve continuously.

# Further reading

If you would like to find out more about energy sources for the future, you might like to read the following books:

Adler, David. *Wonders of Energy*. Mahwah, NJ: Troll Associates, 1983.
Coble, Charles. *A Look Inside Nuclear Energy*. Milwaukee, WI: Raintree, 1983.
Hawkes, Nigel. *Nuclear Power*. New York: Franklin Watts, 1984.
Kaplan, Sheila. *Solar Energy*. Milwaukee, WI: Raintree, 1983.
Metos, Thomas H. and Bitter, Gary G. *Exploring with Solar Energy*. New York: Julian Messner, 1978.
Yates, Madeleine. *Earth Power*. Nashville, TN: Abingdon Press, 1980.

# Index

Aquifers 18
Atom
   structure of 33, 41
   electron 33
   neutron 33, 36, 37
   proton 33

Chemicals
   use of 5, 6, 13

Electricity
   generator 31, 36
   production of 5, 14, 18, 21, 24, 25, 35, 38
Energy
   chemical 13, 45
   heat 8, 12, 13, 17, 18, 27, 31, 35, 36
   kinetic 4, 17, 25, 27
Environment 31, 45

Fossil fuels 5, 45
   coal 5, 6, 17, 43, 45
   gas 5
   oil 5, 17, 43, 45

Fuel rods 35

Geothermal 18, 45
   energy 18, 19, 21
   geysers 17
   hot springs 17

Heat pump 20
Helium 36
Hydroelectric power 25
   dam 25, 26
   Snowy Mountain project 25
Hydrogen 19, 41
   electrolysis 44

Methane 19
Microwaves 15

Nuclear 6, 38, 45
   core 35
   energy 41
   fission 34, 41
   fusion 41
   reactors *see* Reactors
   safety aspects 39–41

Ocean Thermal Energy Converter
   (OTEC) 31, 45
Oil crisis 27

Photosynthesis 7, 44
Plutonium 37
Power stations
   Barstow 11, 15
   geothermal 19
   hydroelectric 25
   Odeilo 9
   solar 16

Radiation 16, 17, 33, 39
Reactors
   Advanced Gas-Cooled 36
   Boiling Water 37
   Fast Breeder 37, 38
   High Temperature Gas-Cooled 36
   Magnox 36

Silicon 14
Solar
   air-tube collector 8
   central receiver collectors 11
   flat mirrors 9
   fresnel lens 9
   panel 7
   parabolic dish reflector 9

photovoltaic cells 14, 15
pond 12
receivers 11, 12, 16

Tidal
   barrage 27
   Bristol oscillating cylinder 28
   energy 27
   Lancaster flexible air bag 30
   oscillating water column 28
   Salter duck 28
   sea clam 30
Tokamak 40
Turbine, steam 5, 30, 31

Uranium 33, 34, 35

Volcanic rock 17, 19

Wave 27, 30, 43
   energy converters 28
Wind 21, 43, 45
   farm 24
   horizontal-axis wind turbine 22
   Musgrove turbine 24
   vertical-axis turbine 23

Wood 6, 44

# Acknowledgments

The publisher would like to thank all those who provided pictures on the following pages: British Aerospace 21; British Telecommunications plc 13; Central Electricity Generating Board 4; Bruce Coleman Limited 17 (John Fennell); Philip Corke 20, 29a, 29b, 31; Courtesy Jet Joint Undertaking 39; Lawrence Livermore Laboratory 40; Maltings Partnership 9, 24, 38, 30; PHOTRI 14; Science Photo Library 2 (A. Hart Davies), 8 (Gazuit), 12, 19 (Soames Summerhays), 22 (Martin Bond), 23 (Tom McHugh), 43; Shell Photographs 5, 6, 7, 41; UKAEA 32, 33, 34, 35, 38; Malcolm S Walker 26, 36, 37; Zefa 10, 11, 25 (Robin Smith).